Learn

Eureka Math®
Grade PK
Modules 3–5

Great Minds® is the creator of *Eureka Math®*, *Wit & Wisdom®*, *Alexandria Plan™*, and *PhD Science®*.

Published by Great Minds PBC.
greatminds.org

Copyright © 2021 Great Minds PBC. All rights reserved. No part of this work may be reproduced or used in any form or by any means—graphic, electronic, or mechanical, including photocopying or information storage and retrieval systems— without written permission from the copyright holder.

ISBN 978-1-63642-664-8

2 3 4 5 6 7 8 9 10 CCR 24 23 22 21

Printed in the USA

A STORY OF UNITS

Mathematics Curriculum

GRADE PK • MODULE 3

Table of Contents
GRADE PK • MODULE 3
Counting to 10

Family Math Newsletters	3
Topic A: *How Many* Questions with up to 7 Objects	7
Topic B: Matching One Numeral with up to 7 Objects	19
Topic C: *How Many* Questions with up to 8 Objects	35
Topic D: Matching One Numeral with up to 8 Objects	45
Topic E: *How Many* Questions with 0 up to 9 Objects	53
Topic F: Matching One Numeral with 0 up to 9 Objects	63
Topic G: *How Many* Questions with up to 10 Objects	73
Topic H: Matching One Numeral with up to 10 Objects	83

A STORY OF UNITS

Mathematics Curriculum

GRADE PK • MODULE 4

Table of Contents
GRADE PK • MODULE 4
Comparison of Length, Weight, Capacity, and Numbers to 5

Family Math Newsletters 101

Topic A: Comparison of Length 105

Topic B: Comparison of Weight 107

Topic D: First and Last 111

Topic E: Are There Enough? 119

Topic F: Comparison of Sets Up to 5 123

Topic G: Comparison of Sets Including Numerals Up to 5 133

End-of-Module Assessment Task 145

A STORY OF UNITS

Mathematics Curriculum

GRADE PK • MODULE 5

Table of Contents

GRADE PK • MODULE 5

Addition and Subtraction Stories and Counting to 20

Family Math Newsletters	149
Topic A: Writing Numerals 0 to 5	153
Topic B: Contextualizing Addition Stories to Solve	169
Topic C: Contextualizing Subtraction Stories to Solve	179
Assessment Template	185
Topic D: Decontextualizing Addition Stories to Solve Using Fingers, Objects, and Drawings	189
Topic E: Decontextualizing Subtraction Stories to Solve Using Fingers, Objects, and Drawings	191
Topic F: Duplicating and Extending Patterns	201
Credits	205
Numeral Cards	207

Grade PK

Module 3

2

A STORY OF UNITS — Family Math Newsletter — PK•3

Grade PK • Module 3 • Topics A–D

Family Math Newsletter

Counting to 10

In the first half of Module 3, students build on their work with numbers to 5 as they explore groups of 6, 7, and 8 objects. Children learn to touch and count up to 8 objects arranged in different ways (e.g., in a straight line or in rows) and extend their ability to make tallies, recognize numerals, and count on their fingers the Math Way (from left to right). Additionally, students strengthen their understanding of *1 more* and discover different ways to take apart numbers (e.g., 7 cubes can be broken up into 5 cubes and 2 cubes).

(Below) Students touch and count objects arranged in lines and circles.

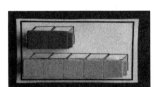

(Above) Students learn to take apart 7 by matching linking cubes to a Partners of 7 Puzzle piece.

Key Standards
- Know number names and the count sequence.
- Count to tell the number of objects.
- Understand that each successive number name refers to a quantity that is 1 larger.

Looking Back
In Module 2, we identified, described, and built shapes.

Looking Ahead
In Topics E–H of Module 3, we will explore 0, 9, and 10. We will spend the most time with 10 since it is foundational to understanding place value.

Words and Key Terms

Terminology
- Counting the Math Way
- Eight (8)
- How many?
- Number path
- One more/larger
- One less/smaller
- Pair
- Seven (7)
- Six (6)
- Tally mark

How to Help at Home

- Touch and count up to 8 objects together. During playtime, count up to 8 toy cars. Move the cars into a line or a circle and count again.
- Buy or make a set of numerals from 1 to 8 (paper, foam, or magnets work well). Show a number on your fingers. Ask, "Which number shows how many fingers I am holding up?" Switch roles and let your child show a number on his fingers.
- Ask for help with counting during everyday experiences. While cooking, say, "I need 6 mushrooms. Can you count out 6 mushrooms for me?"
- Continue to sing songs that involve counting forward or back, such as "The Ants Go Marching," "This Old Man," "Eight Little Ducks Went Out to Play," or "Eight Little Monkeys Jumping on the Bed."
- Some lessons require additional materials. See your student's teacher for the additional materials required for each lesson.

Module 3: Counting to 10

Copyright 2021 © Great Minds PBC

A STORY OF UNITS | Family Math Newsletter | PK•3

Spotlight on Math Models

Children will use key mathematical models throughout their elementary years. One of these models is the 5-group, a tool Pre-Kindergarten students will use to show and work with numbers 1–10.

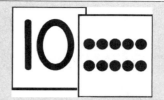

Sample Chant
(from Module 3, Lesson 12)

One Potato, Two Potato

Students say the "One Potato, Two Potato" rhyme to help their teacher count the potatoes she will slice to make french fries:

One potato, two potato,

Three potato, four,

Five potato, six potato,

Seven potato, more.

The teacher then asks, "What does *more* mean?" She adds another potato to the group and says, "What is 7 and *1 more*? Let's count!"

This task reviews counting up to 7 objects, as well as the concept of 1 more. In the lesson that follows, students build on this understanding to relate 7 and 1 more to 8.

5-Groups

Five is a key number in helping children understand 6, 7, 8, 9, and 10. 5-groups organize these numbers as 5 and some more (e.g., 6 is 5 and 1 more, or 5 + 1). One easy way to see this relationship is with dots lined up in groups of 5 as pictured below. These make it easy to see each number in relation to 5 (5 + 1, 5 + 2, 5 + 3, 5 + 4, 5 + 5). Without experience with 5-groups, children have little understanding of numbers 6–10 other than a general sense that the numbers are getting larger.

Why is this important? The patterns that you see in the dot cards above can be used as tools for solving addition and subtraction problems in Kindergarten and Grade 1. For example, you can easily see that 8 − 3 = 5 and 8 − 5 = 3. You can also see that 8 needs 2 more to make 10.

Dots are not the only way to show 5-group formations. Fingers clearly show the relationship between 5 and the numbers 6–10 (5 fingers on one hand and some more fingers on the other hand). A color change at 5, or organization of objects or drawings in groups of 5, can also help children see this important relationship.

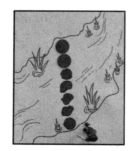  

4   Module 3:   Counting to 10   EUREKA MATH

A STORY OF UNITS — Family Math Newsletter PK•3

Grade PK • Module 3 • Topics E–H
Family Math Newsletter

Counting to 10

In the second half of Module 3, students build on their previous number work as they explore groups of 0, 9, and 10 objects. More time is spent with 10, since it is important for understanding place value in later grades. Children learn to touch and count up to 10 objects arranged in different ways (e.g., in a straight line or in a circle) and extend their ability to make tallies, recognize numerals, and count on their fingers the Math Way (from left to right). Students strengthen their understanding of *1 more* and discover different ways to take apart numbers (e.g., 10 cubes can be broken up into 9 cubes and 1 cube).

At the Pollen Café, students place bee customers in their seats on the number path and learn to identify the total number of bees both by counting and using the corresponding numerals on the number path. The teacher models how to make tallies before students practice on their own.

How to Help at Home

- Touch and count 10 objects together. During playtime, count up to 10 blocks, lining them up as you count. Stack the blocks and count again. The number of blocks stays the same!
- Look for numerals when walking, driving, or taking the bus. "I see the number 10. Let's clap 10 times!"
- At snack time, line up 10 blueberries and ask your child to count them. Each time she eats a berry, have her touch and count how many are left.
- Use the illustrations in picture books to count as you read together. "I wonder how many vegetables Mr. McGregor planted in his garden. Let's count them!"
- Share information about your child's counting with the teacher. If you notice that your child is skipping a number while counting, communicate that in a note to the teacher.
- Some lessons require additional materials. See your student's teacher for the additional materials required for each lesson.

Key Standards
- Know number names and the count sequence.
- Count to tell the number of objects.
- Understand that each successive number name refers to a quantity that is 1 larger.

Looking Back
In the first half of Module 3, we expanded on our work with numbers to 5 to explore 6, 7, and 8.

Looking Ahead
In Module 4, students will learn to compare as they explore length, weight, and capacity. They will also strengthen their understanding of numbers as they compare sets of up to 5 objects.

Words and Key Terms

Terminology
- Counting the Math Way
- Counting in a circle
- Counting in a line
- Counting in rows
- How many?
- Nine (9)
- Number path
- One more/larger
- One less/smaller
- Put together
- Take apart
- Tally mark
- Ten (10)
- Zero (0)

Module 3: Counting to 10

A STORY OF UNITS — Family Math Newsletter — PK•3

Spotlight on Math Models

Children will use key mathematical models throughout their elementary years. One of these models is the number path, a tool Pre-K students will use to connect counting and numbers 1–10.

| 1 | 2 | 3 | 4 | 5 | 6 | 7 | 8 | 9 | 10 |

Sample Activity
(from Module 3, Lesson 29)

Pollen Café

As waiters at the Pollen Café, students place bee customers in special seats on the number path. Then, they tally the orders for flowers.

Children bring back the correct number of flowers and give each bee a snack.

This task reviews counting up to 9 objects and introduces tallying to 9. The number path supports children in counting and matching their count to a numeral.

Number Path

The number path is used in Pre-K, Kindergarten, and Grade 1 to help children work with numbers and visualize the number sequence. The number path starts at 1 and has a single space for each number. A color change at 5 shows the relationship between 5 and the numbers 6, 7, 8, 9, and 10.

Starting in Module 1, children began to see the number path to 5 using stairs that show the total number at each step (as pictured on right). In this module, children continue to build the number path to 10, noticing that each number in the sequence is 1 larger.

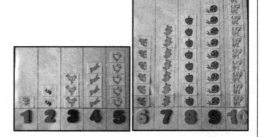

In Module 3, children work with the path in a new way, placing one object in each space on the number path. Children also see that the last number said (and the last space filled) tells the number of objects counted. For example, in the image of the sheep below, the student can touch and count the sheep and come to understand that the numeral 6 tells the total number of sheep.

In Kindergarten and Grade 1, students will learn to use the number path to solve addition and subtraction problems.

Module 3: Counting to 10

Name _____   Date _____

Draw 1 more soccer ball in each line.

8

creek mat

Lesson 2: Use linear configurations to count 6 and 7 in relation to 5.

Name _____ Date _____

Draw 6 eggs. Draw lines to show the 6 chicks standing.

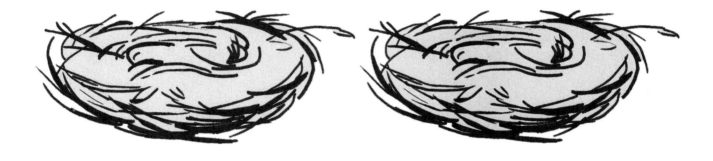

Lesson 3: Count to 6 and 7 left to right with fingers.

A STORY OF UNITS — Lesson 3 Template — PK•3

creek mat

Lesson 3: Count to 6 and 7 left to right with fingers.

Copyright 2021 © Great Minds PBC

13

14

Name _____ Date _____

Draw 7 eggs. Draw lines to show the 7 chicks standing.

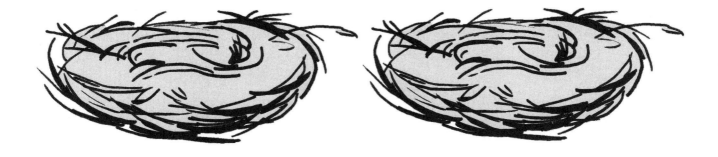

Name _____   Date _____

Color the socks to show pairs.

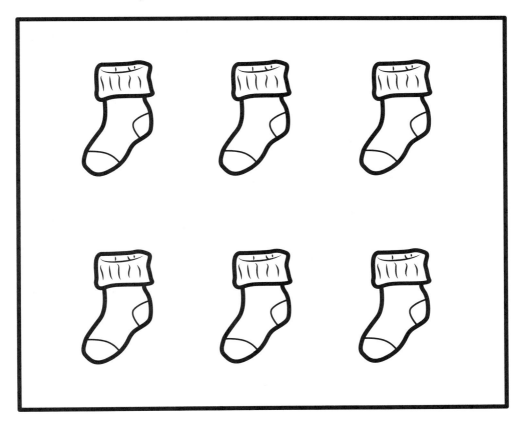

Color the set below that has 6.

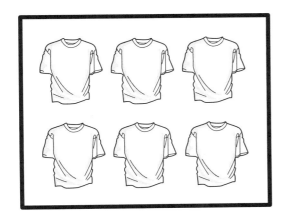   

A STORY OF UNITS

Lesson 6 Template 1 PK•3

Cut along dashed lines to prepare the puzzles.

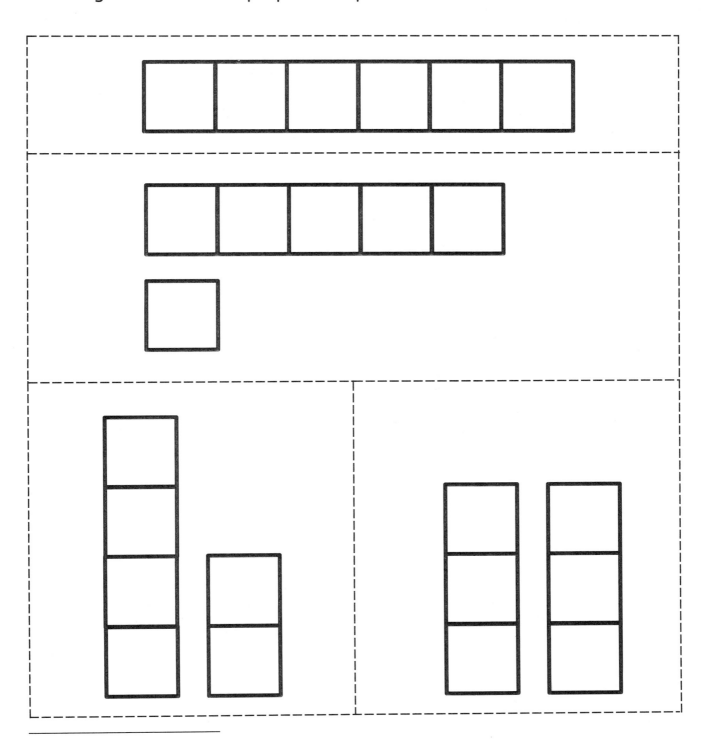

partners of 6 puzzles

Lesson 6: Compose 6, and then decompose into two parts. Match to the numeral 6.

A STORY OF UNITS
Lesson 7 Template PK•3

Cut along dashed lines to prepare the puzzles.

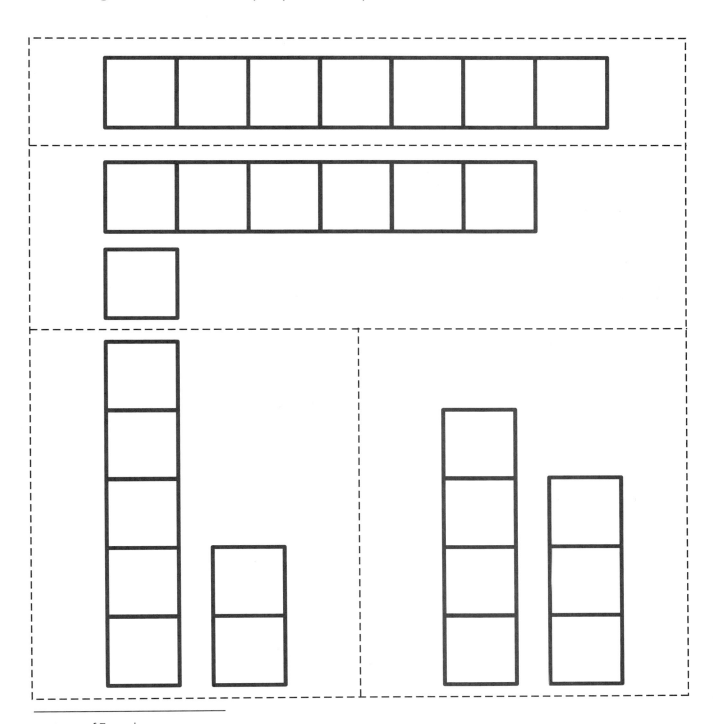

partners of 7 puzzles

Lesson 7: Compose 7, and then decompose into two parts. Match to the numeral 7.

22

A STORY OF UNITS **Lesson 8 Template** PK•3

Cut along dashed lines to prepare the cards.

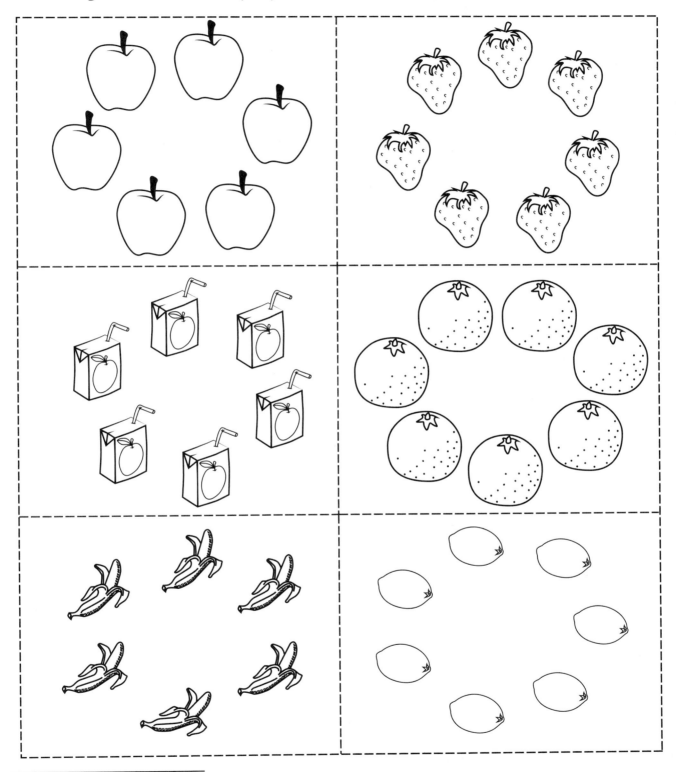

circular configuration cards

Lesson 8: Count 6 and 7 objects in circular configurations.

underwater mat

26

A STORY OF UNITS Lesson 9 Template 2 PK•3

Cut along dashed lines to prepare the cards.

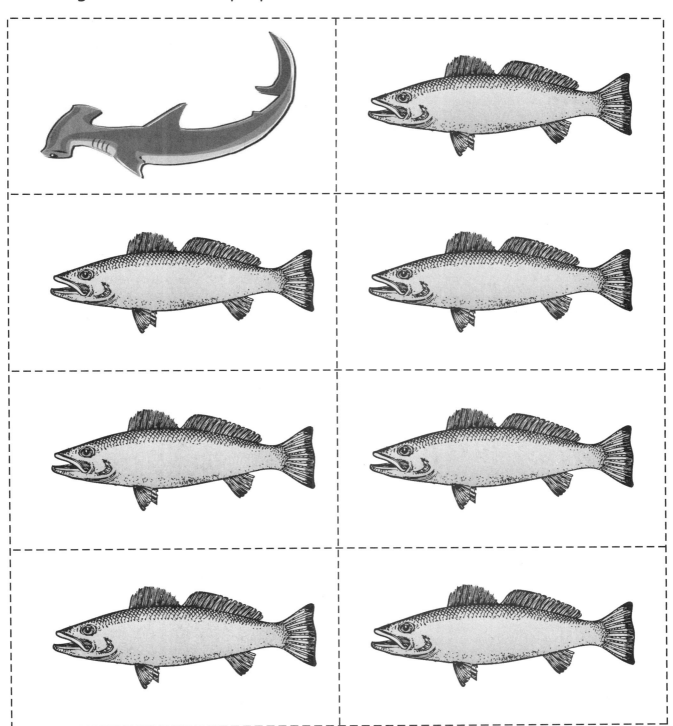

shark and fish

Lesson 9: Arrange and count 6 and 7 objects in varied configurations.

27

28

Cut along dashed lines to prepare the cards.

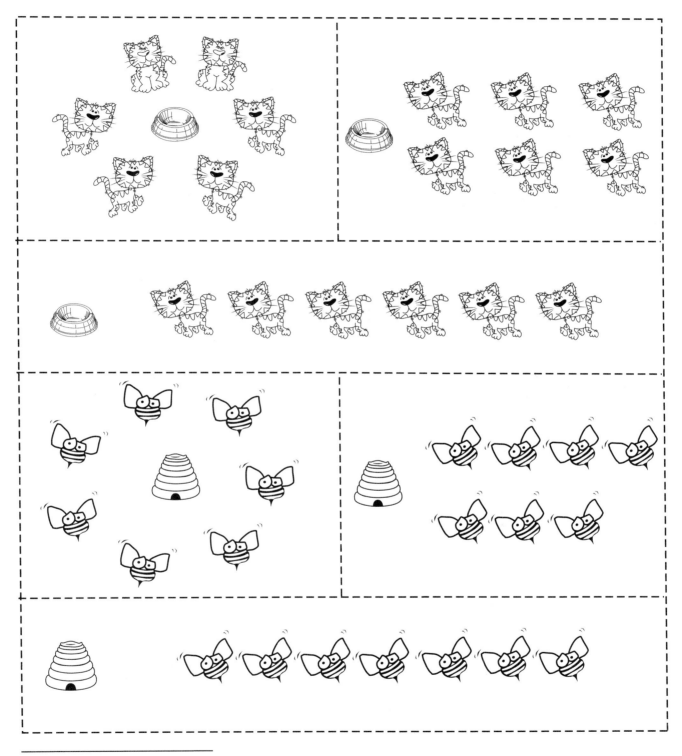

6–7 picture cards

Lesson 9: Arrange and count 6 and 7 objects in varied configurations.

30

A STORY OF UNITS

Lesson 10 Problem Set PK•3

Name _____ Date _____

Lesson 10: Tally 6 and 7 objects.

31

A STORY OF UNITS　　　　　　　　　　　　　　　　　Lesson 10 Template　PK•3

Cut out each number path. Tape the number paths together to create one sequence from 1–10.

1		2		3		4		5	

6		7		8		9		10	

number path

Lesson 10:　Tally 6 and 7 objects.　　　　　33

Copyright 2021 © Great Minds PBC

Save this item for future use.

Ollie Octopus

36

A STORY OF UNITS — Lesson 13 Template — PK•3

creek mat

Lesson 13: Use linear configurations to count 8 in relation to 5.

38

Name _____ Date _____

Draw 8 eggs. Draw lines to show the 8 chicks standing.

A STORY OF UNITS

Lesson 15 Problem Set PK•3

Name _____ Date _____

Trace and count the legs on Ansel Ant and Spencer Spider.

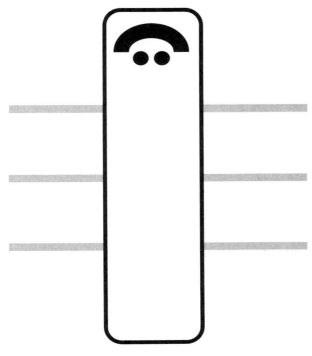

42

| A STORY OF UNITS | Lesson 15 Template | PK•3 |

Cut along the lines to separate the socks.

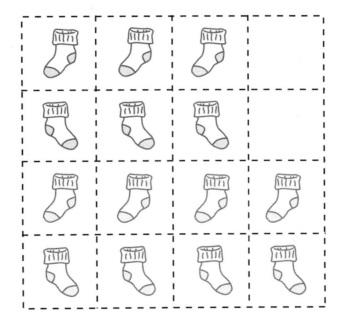

_____

socks for Ansel and Spencer

Lesson 15: Count 8 objects in array configurations.

Copyright 2021 © Great Minds PBC

43

A STORY OF UNITS Lesson 16 Template PK•3

Cut along dashed lines to prepare the puzzles.

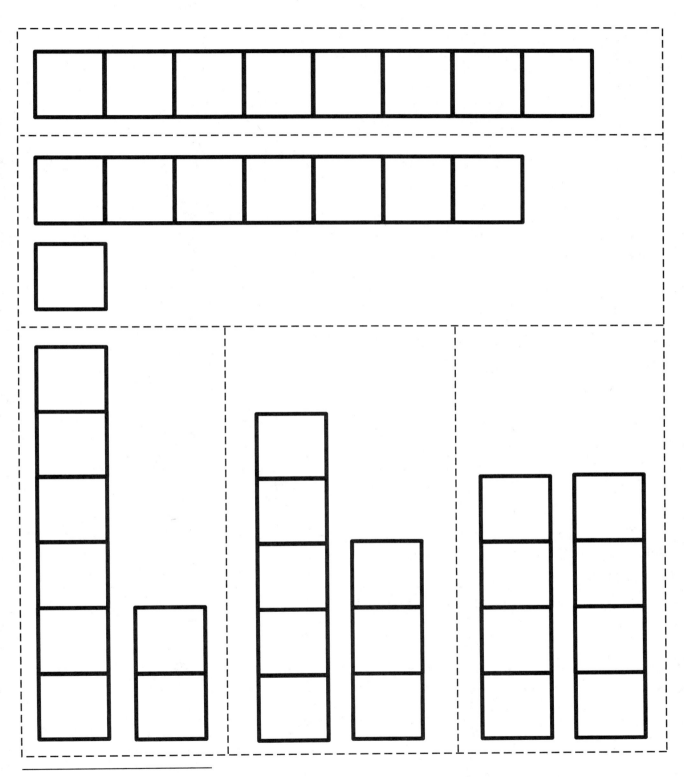

partners of 8 puzzles

Lesson 16: Compose 8, and decompose into two parts. Match to the numeral 8.

45

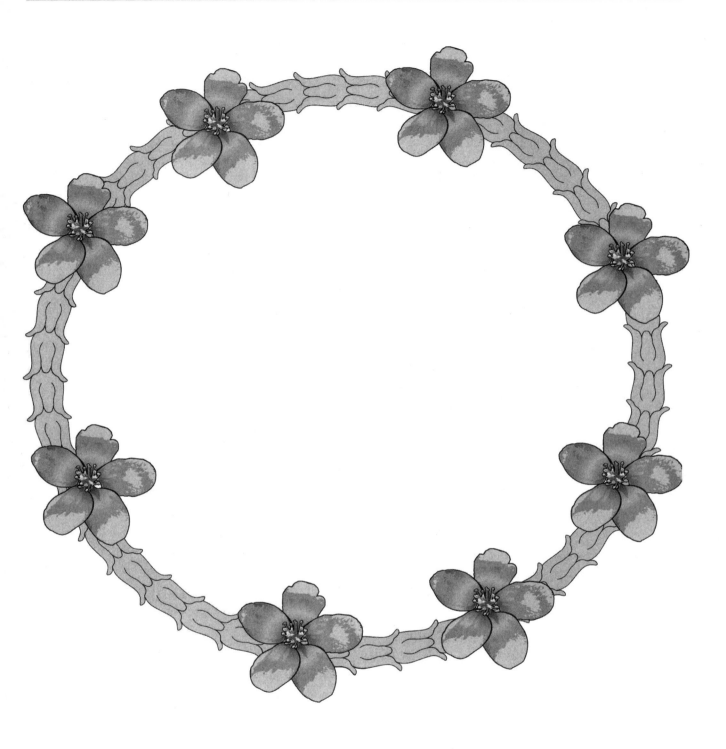

crown picture

48

A STORY OF UNITS — Lesson 17 Template 2 — PK•3

Cut along dashed lines to prepare the cards.

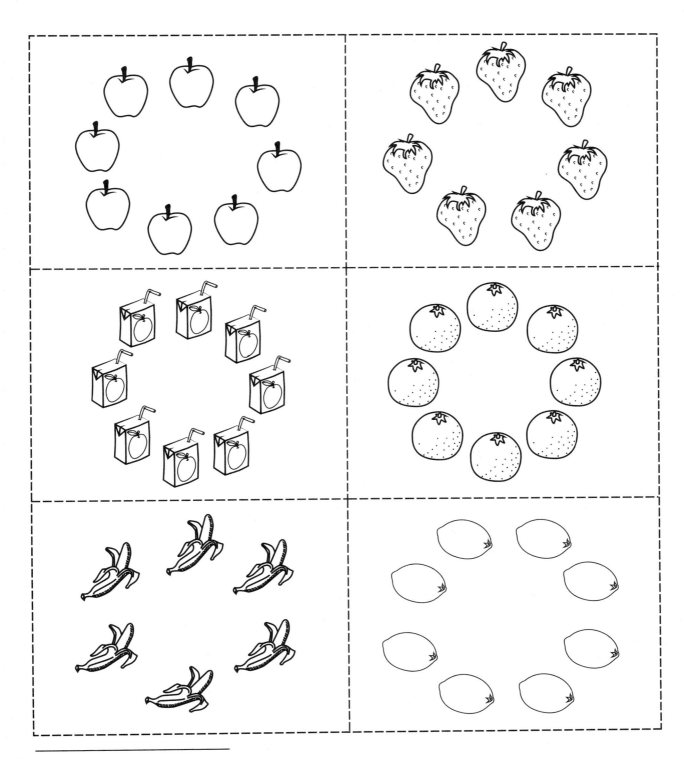

circular configuration cards

Lesson 17: Count 8 objects in circular configurations.

50

A STORY OF UNITS Lesson 19 Template PK•3

Cut out each number path. Tape the number paths together to create one sequence from 1–10.

number path

Lesson 19:   Tally 8 objects.

51

52

A STORY OF UNITS — Lesson 21 Template — PK•3

underwater mat

Lesson 21: Introduce zero.

53

54

creek mat

56

Name _____ Date _____

Draw 9 eggs. Draw lines to show the 9 chicks standing.

58

Three Blind Mice

Three blind mice. Three blind mice.

See how they run. See how they run.

They all ran after the farmer's wife, who chased them away with a fork and a knife.

Did you ever see such a sight in your life,

As three blind mice?

Three Little Kittens

Three little kittens they lost their mittens,

And they began to cry,

Oh, mother dear, we sadly fear

Our mittens we have lost.

What! Lost your mittens, you naughty kittens!

Then you shall have no pie.

Mee-ow, mee-ow, mee-ow.

No, you shall have no pie.

---

nursery rhymes

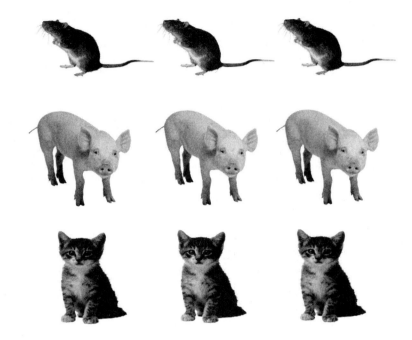

animal array

62

Cut along dashed lines to prepare the puzzles.

partners of 9 puzzles

64

Lesson 26 Template 2 PK•3

Cut along dashed lines to prepare the puzzles.

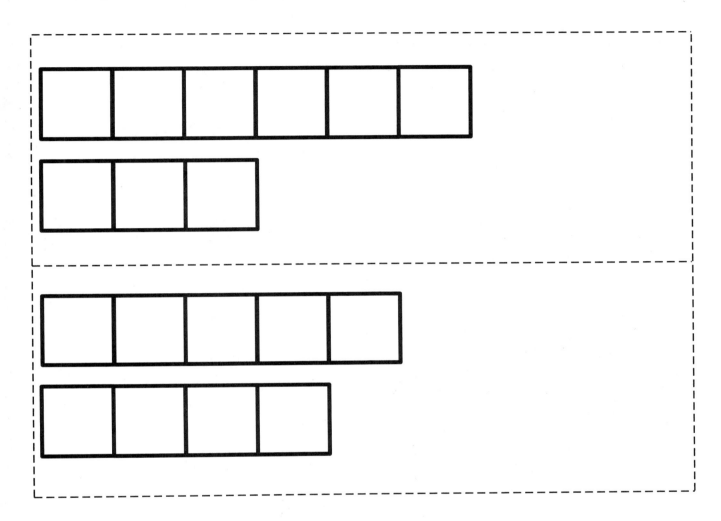

partners of 9 puzzles

Lesson 26: Compose 9, and decompose into two parts. Match numerals 0 and 9 to no objects and 9 objects.

66

A STORY OF UNITS   Lesson 27 Template 1   PK•3

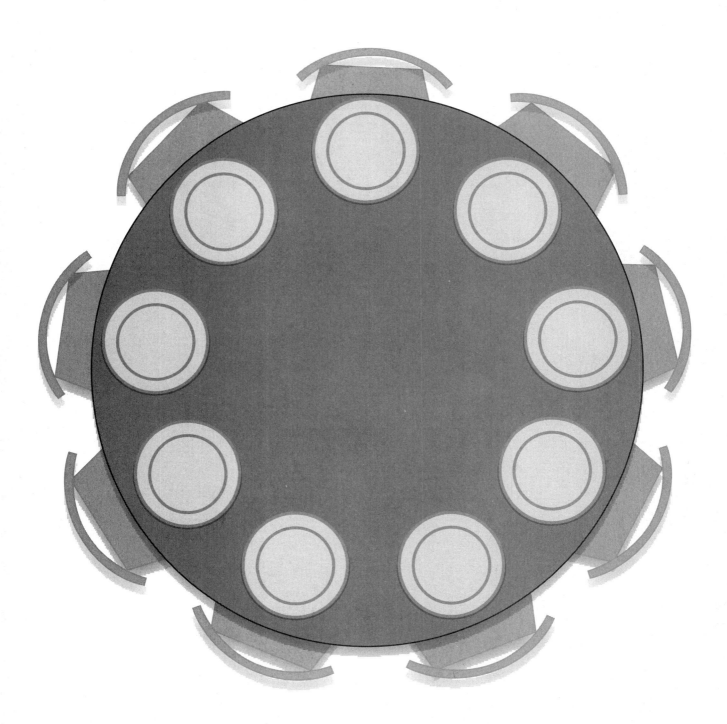

table template

Lesson 27: Count 9 objects in circular configurations.

Cut along dashed lines to prepare the cards.

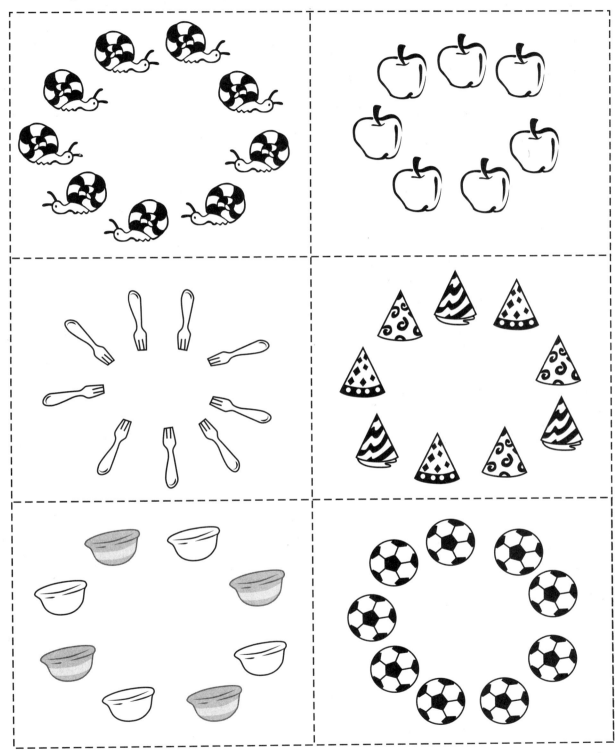

circular configuration cards

Lesson 27: Count 9 objects in circular configurations.

70

Cut out each number path. Tape the number paths together to create one sequence from 1–10.

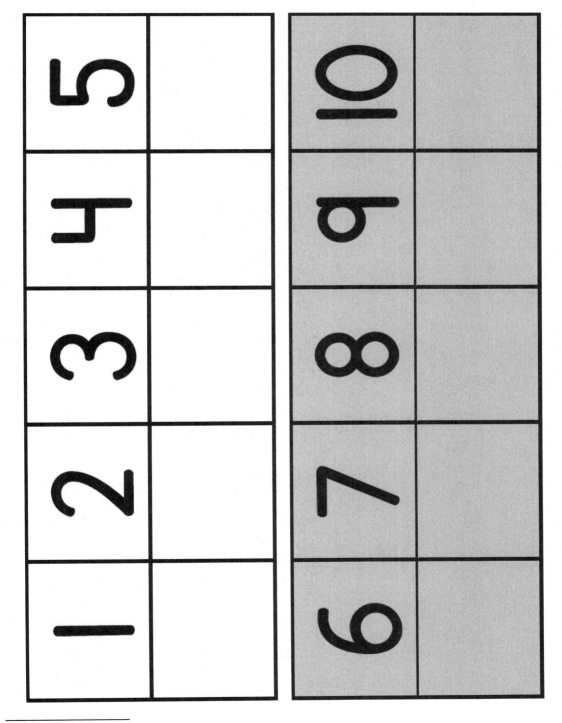

number path

Lesson 29: Tally 9 objects.

71

A STORY OF UNITS • Lesson 31 Template • PK•3

Cut along dashed lines to separate the images.

orange slices

Lesson 31: Introduce 10, and relate 10 to 9 and *1 more*.

73

74

creek mat

Name _____ Date _____

Draw 10 eggs. Draw lines to show the 10 chicks standing.

A STORY OF UNITS Lesson 34 Template PK•3

Cut along dashed lines to prepare the cards.

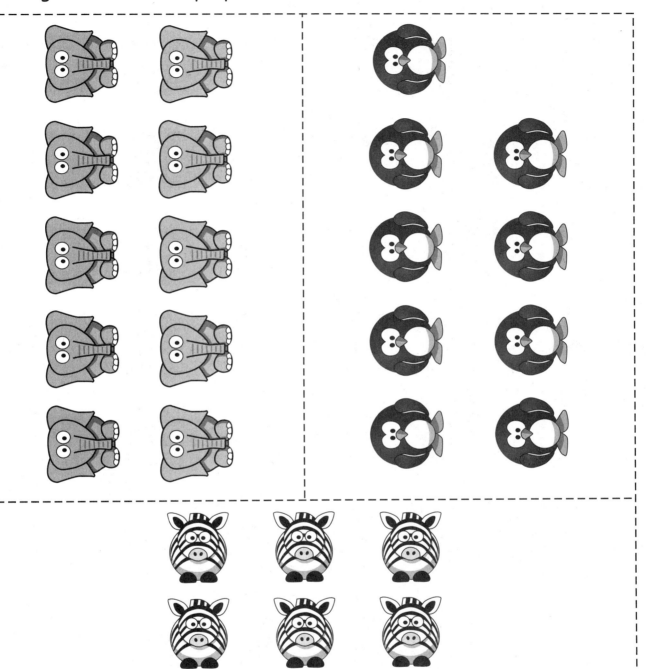

animal array cards

Lesson 34: Count 10 objects in array configurations.

79

80

A STORY OF UNITS

Lesson 34 Template   PK•3

Cut along dashed lines to prepare the cards.

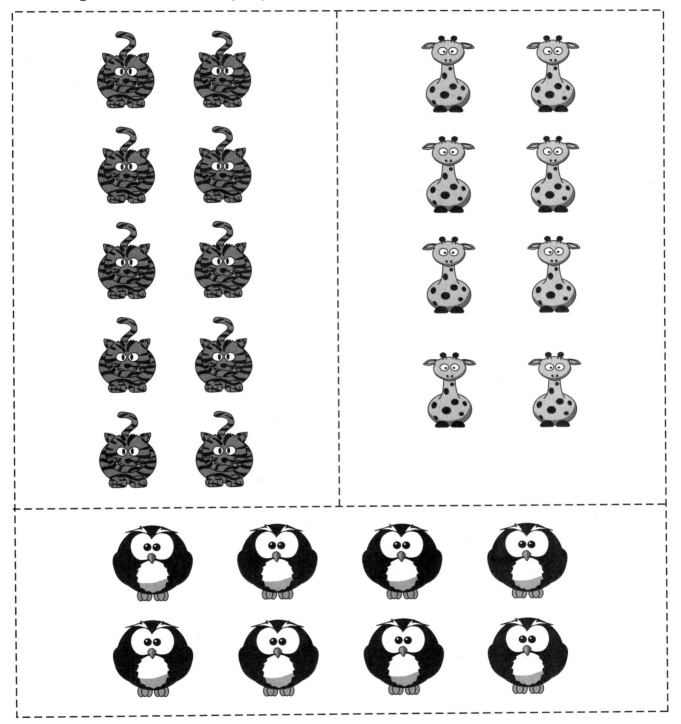

animal array cards

Lesson 34:   Count 10 objects in array configurations.

81

82

A STORY OF UNITS Lesson 35 Template 1 PK•3

Cut along dashed lines to prepare the puzzles.

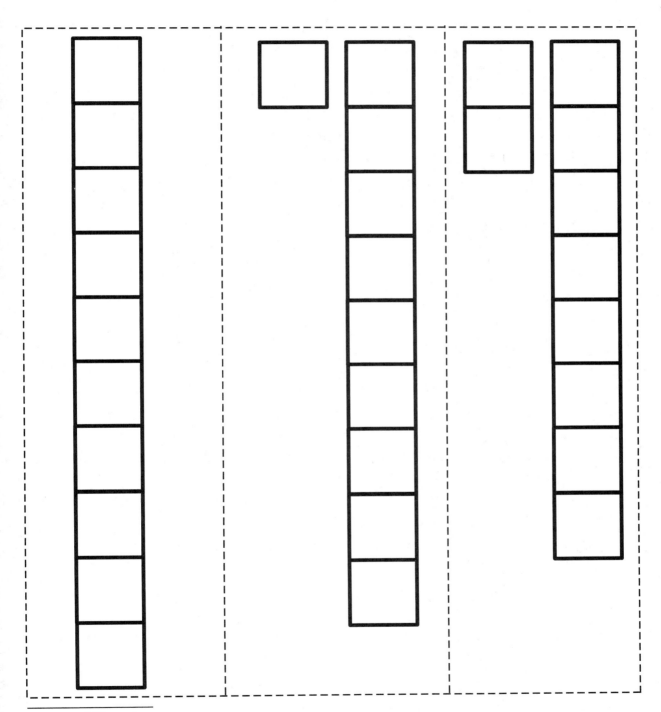

partners of 10 puzzles

Lesson 35: Compose 10, and decompose into two parts. Match to the numeral 10.

83

84

A STORY OF UNITS

Lesson 35 Template 2 PK•3

Cut along dashed lines to prepare the puzzles.

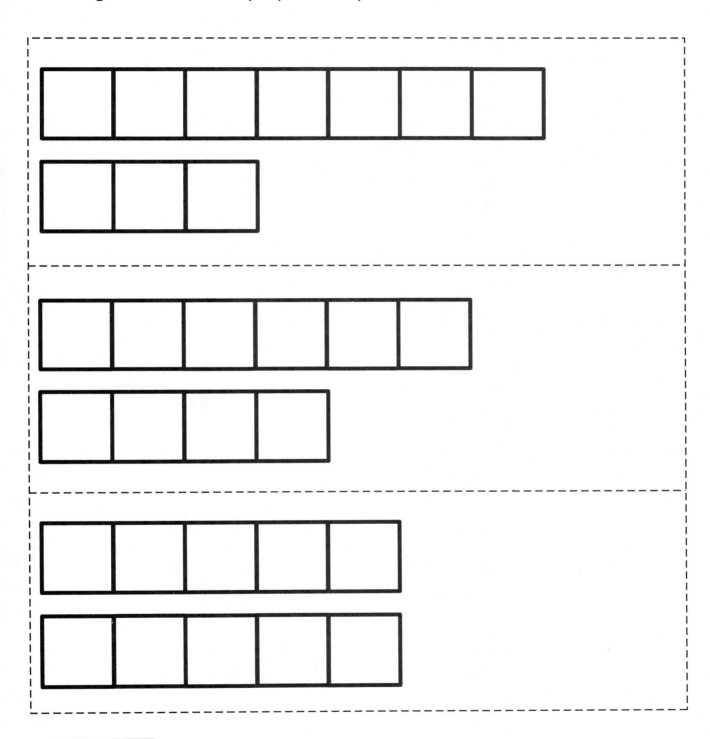

partners of 10 puzzles

Lesson 35: Compose 10, and decompose into two parts. Match to the numeral 10.

86

decomposition mat

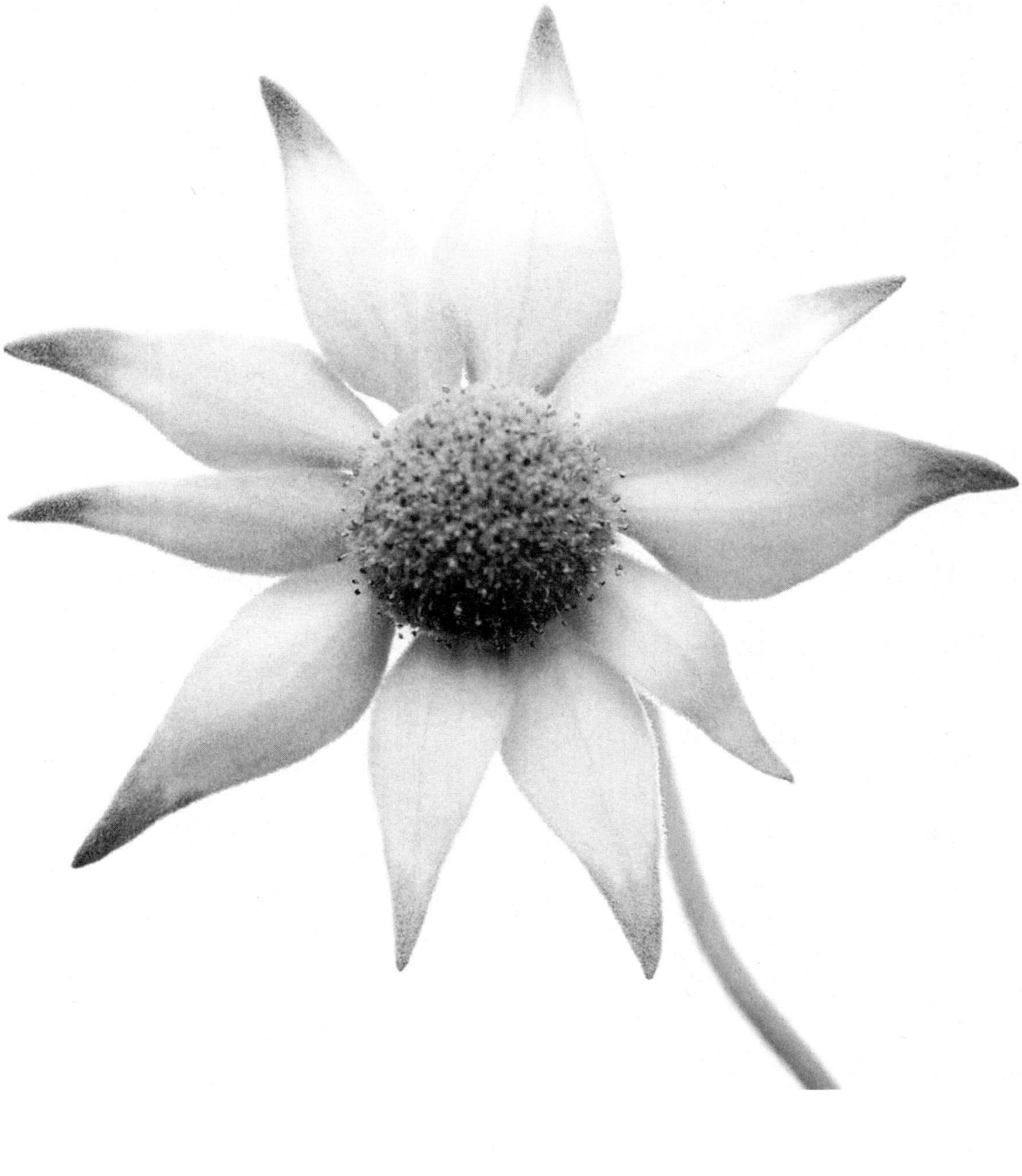

flower image

90

A STORY OF UNITS

Lesson 37 Template 2   PK•3

Cut along dashed lines to prepare the cards.

circular configuration cards

Lesson 37: Arrange and count 10 objects in circular configurations.

91

Cut out each number path. Tape the number paths together to create one sequence from 1–10.

5		10	
4		9	
3		8	
2		7	
1		6	

number path

Lesson 40: Tally 10 objects.

A STORY OF UNITS
Lesson 42 Template PK•3

Cut along dashed lines to prepare the cards.

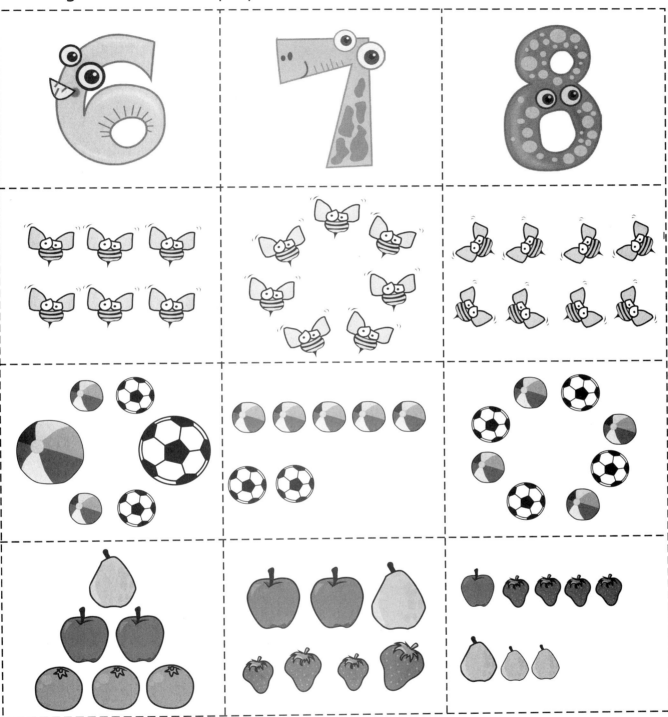

---

pictures and shapes

Lesson 42: Culminating Task—represent numbers 6–10 using objects, images, and numerals in a number book.

95

96

A STORY OF UNITS Lesson 42 Template PK•3

Cut along dashed lines to prepare the cards.

pictures and shapes

Lesson 42: Culminating Task—represent numbers 6–10 using objects, images, and numerals in a number book.

98

Grade PK
Module 4

100

A STORY OF UNITS   **Family Math Newsletter**   PK•4

Grade PK • Module 4 • Topics A–C
Family Math Newsletter

Comparison of Length, Weight, Capacity, and Numbers to 5

In Topics A–C of Module 4, students compare and explore lenghts, weights, and capacities. For example, students learn to line up the endpoints when comparing length, use a balance scale to compare weight, and pour sand into containers of different sizes and shapes as they compare capacity.

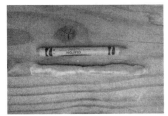

(Above) Children make clay snakes that are longer than a crayon.

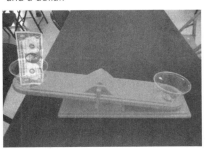

(Below) Students use a balance scale to compare the weight of a quarter and a dollar.

Key Standards
- Describe and compare measurable attributes of length, weight, and volume.
- Compare numbers.
- Identify *first* and *last* related to order or position.

Looking Back
In Module 3, students used what they learned about numbers to 5 to explore numbers 6–10 and 0. Using 5 as a starting point, they learned that 6 is one more than 5. They counted up to 10 objects in different arrangements, made tallies, and learned to recognize numerals to 10. Students also practiced counting *1 more* and explored different ways to take apart numbers.

Looking Ahead
In Topics D–G of Module 4, students identify first and last and compare sets of up to 5 objects.

How to Help at Home
- Play I Spy while walking or driving. Look for objects of different length or weight. "I spy something heavy and round." (A rock!) "I spy something tall and green." (A tree!) "I spy something shorter than a foot." (A worm!)
- Use blocks to build *trains* of different lengths. Ask, "Which train is shorter?" Say, "Let's try to build another train the same length!"
- Explore capacity with water play. Set out container of different sizes and shapes. Ask, "Which container do you think can hold the most water?" or "Do you think all the water in your cup will fit in this bowl?" Pour water back and forth among the containers and make observations.
- Continue to find opportunities to count in everyday experiences. "I wonder how many steps there are. Let's count as we walk up them!"
- Some lessons require additional materials. See your student's teacher for the additional materials required for each lesson.

Suggested Words and Key Terms
- About the same as
- Are there enough?
- Balance scale
- Big/small
- Compare
- Exactly
- Extra
- Fewer
- First/last
- Greater than/less than
- Heavy/light
- Heavier than/lighter than
- Length
- Less
- Longer than/shorter than
- More
- More than/less than
- Tall/short

Spotlight on Math Models

Children will use key mathematical models throughout their elementary years. One of these models is the linking cube tower/train, a tool Pre-K students will use to compare length and numbers.

Sample Counting Vignette
(From Module 4, Lesson 9)

Ice Cream

Teacher: I'm going to count and make a mistake on purpose. Instead of saying a number, I'll say "ice cream!" Isn't that silly? Listen closely and see if you can tell what number I should've said.

1, 2, ice cream!

Students: 3.

Teacher: Very good. Listen again: 1, 2, 3, 4, ice cream!

Students: 5.

Teacher: Excellent. This one will be a bit of a challenge. Ready? 1, 2, ice cream, 4, 5.

At this point in the year, students are steadily gaining mastery of the counting sequence. This activity challenges them to detect an error in the familiar order of numbers. Teachers work within a range that is comfortable for all students, and slowly build up.

Linking Cube Tower and Linking Cube Train

The linking cube tower and train are powerful tools that are used through Grade 2. In the first half of Module 4, students simply hold the towers (linking cubes situated vertically) next to each other to make *longer than, shorter than,* and *same as* statements. Informally, students notice that each tower is built from equal units, an important measurement concept setting the foundation for the ruler, number line, and fractions.

Toward the end of this module, students use the linking cube trains (linking cubes situated horizontally) to compare numbers. They count the cubes and then build each train. From their work comparing towers and trains, they can say, "5 is more than 3."

Because young children commonly use the words *big* and *small* to describe most objects, this module focuses on teaching students vocabulary that allows them to be more precise in their description of objects. Learning and using comparative statements like *longer* or *shorter than, heavier* or *lighter than,* and *more* or *less than* with objects is a bridge to comparison of numbers—*greater* or *less than.* Using number towers and trains is one of the first concrete steps in this process.

Say Ten Counting

At this time of year, students are ready to count and think about numbers above 10 in a way that highlights place value. In this module, students practice counting the *Say Ten* way, "ten one (11), ten two (12), ten three (13), ten four (14)", and so on. *Say Ten* counting is taught in playful ways to expose students to the notion that the number eleven has a ten and a 1 inside it. They continue practice counting and saying numbers the regular way as well.

A STORY OF UNITS — **Family Math Newsletter** PK•4

Grade PK • Module 4 • Topics D–G

Family Math Newsletter

Comparison of Length, Weight, Capacity, and Numbers to 5

The second half of Module 4 begins with an exploration of *first* and *last* when objects are counted in linear, array, circular, and scattered arrangements. Students use the language of comparison they began to develop when working with length, weight, and capacity as they compare sets of up to 5 objects. This module culminates with students counting to compare sets of objects, "4 cats is more than 3 cats," and finally, numbers, "4 is greater than 3."

Key Standards
- Compare numbers.
- Identify *first* and *last* related to order or position.

Looking Back
In the first half of Module 4, students compared and explored length, weight, and capacity.

Looking Ahead
In Module 5, students learn to write numbers to 5, explore addition and subtraction stories, and count to 20.

(Above) Students compare game pieces.

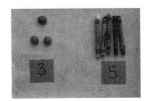

(Below) Attaching numbers to sets gradually leads students to compare numbers alone.

Suggested Words and Key Terms
- Are there enough?
- Compare
- Equal to
- Exactly enough
- Exactly the same
- Extra
- Fewer
- Fewer than
- First
- Greater
- Greater than
- How many?
- Last
- Less
- Less than
- Match
- More
- More than
- Set
- The same as

How to Help at Home

- Before counting some tomatoes with your child, decide which tomato to count first and which to count last. After counting, make a new decision and see that the count is the same!
- Count toys and compare sets during play. Ask, "How many cars do you have?" "How many trucks?" "I wonder if you have fewer cars or trucks. Let's line them up and see!"
- When walking, make comparison statements, "My steps are longer than your steps." "I take fewer steps than you to go places." "I took 4 steps, and you took 5 steps." "4 is less than 5."
- Read counting books or recite nursery rhymes and encourage the child to count images. By the end of Pre-Kindergarten, students should be able to count to 20 by rote (on their own), but if they can touch and count to 20, that's terrific!
- Some lessons require additional materials. See your student's teacher for the additional materials required for each lesson.

A STORY OF UNITS — Family Math Newsletter PK•4

Spotlight on Math Vocabulary

Children use key mathematical vocabulary throughout their elementary years. The language of comparison (greater than and less than) is vocabulary Pre-K students use to compare numbers.

Sample Activity
(From Module 4, Lesson 20)

Clay Numeral 2

Teacher: Take your clay and roll it into a long, skinny, snake.

Students: (Manipulate clay.)

Teacher: Put your snake on the 2, starting at the star.

Students: (Use their clay to first make the curved part of the 2 and then the straight part.)

Teacher: If you finish early, use your finger to trace the 2, starting at the star.

This activity anticipates writing numerals in Module 5 and is intended to familiarize students with correct numeral formation. In addition, students use their fine motor skills to manipulate the clay.

A Focus on Models of Comparison

In the first half of this module, children compared length, weight, and capacity. Now, they transition into comparing numbers by matching two groups of objects and considering if there are *enough, not enough*, or *more than enough*. For example, the model below shows the following statement to be true, "There are not enough crayons for each paper."

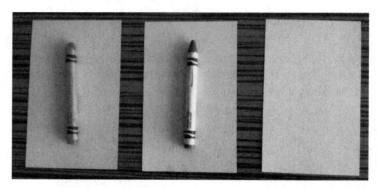

The next step in comparing numbers is to match the objects in each group to find out if there are "*more* crayons *than* papers," "*fewer* crayons *than* papers," or "the *same* number of crayons *as* papers." At this point, the students are counting and saying the number of each group but making their comparison statements as they match the two groups of objects.

In the culminating lessons of this module, students attach a number card to each group of objects to make *greater than* or *less than* statements. For example, "3 is greater than 2" or "2 is less than 3." Finally, students are shown a pair of number cards (up to 5) and are asked to make *greater than* or *less than* statements without objects. Then, they verify their statements by making linking cube towers/trains.

Working to compare by using the abstract (number cards) and the concrete (linking cube towers) develops students' number sense as they relate numbers to each other. This point is emphasized, so when students work on comparison, they can answer questions such as, "How many *more* apples does Maria have *than* Armen?" A solid foundation is being carefully laid right now! Together, everyone can!

creek mat

Cut along dashed lines to prepare the cards.

weight collage

Name _____ Date _____

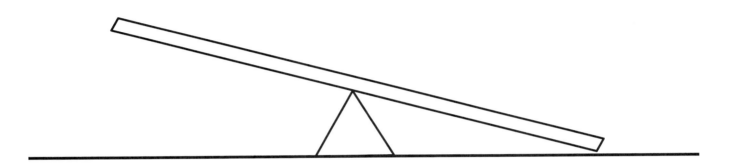

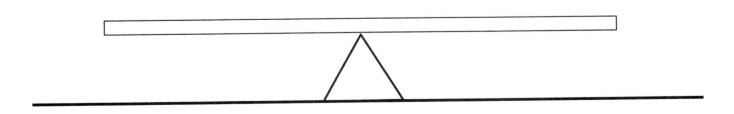

Lesson 8: Compare weight using *heavier than*, *lighter than*, and *the same as* with balance scales.

A STORY OF UNITS　　　　　　　　　　　　　Lesson 13 Problem Set　PK•4

Name _____　Date _____

Lesson 13:　Identify first and last in a scattered configuration with 2–5 objects.　111

Copyright 2021 © Great Minds PBC

112

canoe

Lesson 14: Identify first and last in a linear configuration with 2–10 objects.

114

rowboat

116

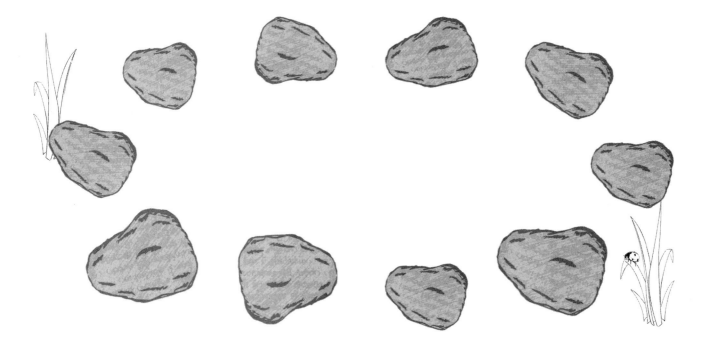

circle of rocks

118

chairs

Cut along dashed lines to prepare the card.

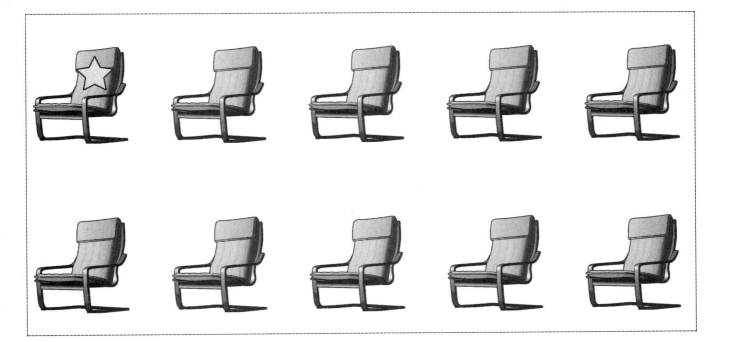

movie theater chairs

Lesson 18: Compare: Match to find there are enough, with some extras.

Cut along dashed lines to prepare the cards.

numeral formation cards

124

A STORY OF UNITS

Lesson 19 Template   PK•4

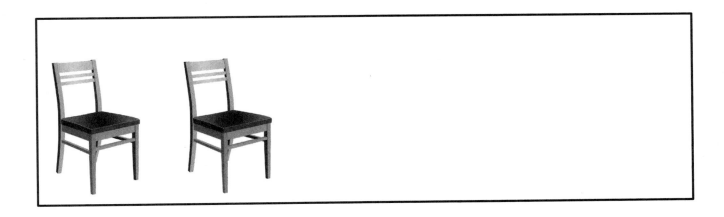

chairs

Lesson 19: Count and match to compare using *fewer than* statements.

125

A STORY OF UNITS　　　　　　　　　　　　　　Lesson 20 Problem Set　PK•4

Name _____ Date _____

(Fold on the line.)

Lesson 20: Count and match to compare using *the same as* statements.

128

Cut along dashed lines to prepare the card.

numeral formation cards

130

Cut along dashed lines to prepare the card.

---
numeral formation cards

132

A STORY OF UNITS Lesson 23 Template 1 PK•4

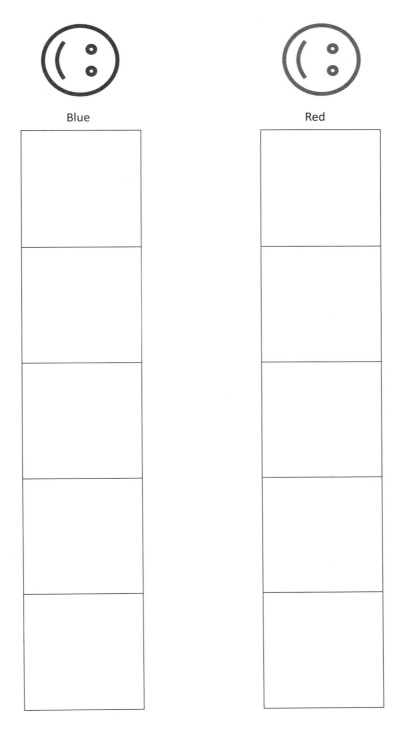

Blue Red

baseball scoreboard

Lesson 23: Compare a number of objects using *more than* or *the same as* statements. 133

Copyright 2021 © a Minds PBC

134

A STORY OF UNITS

Lesson 23 Template 2   PK•4

Cut along dashed lines to prepare the pictures.

small baseball cutouts

Lesson 23: Compare a number of objects using *more than* or *the same as* statements.

135

Cut along dashed lines to prepare the card.

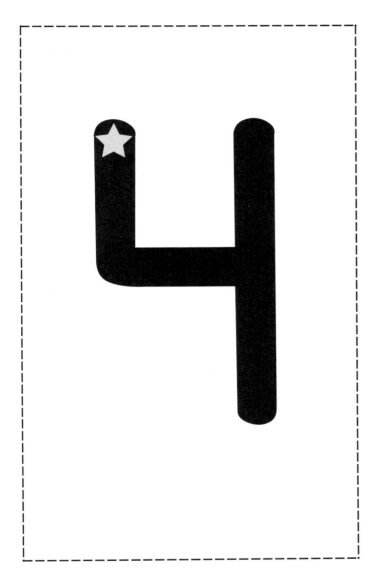

numeral formation cards

138

grass and water

140

Cut along dashed lines to prepare the card.

---
numeral formation cards

142

A STORY OF UNITS   Lesson 27 Fluency Template   PK•4

stars and stripes

Lesson 27: Count and match to compare two sets of linking cube towers.

144

5-box template

146

Grade PK
Module 5

148

| A STORY OF UNITS | Family Math Newsletter |

Grade PK • Module 5 • Topics A–C
Family Math Newsletter

Addition and Subtraction Stories and Counting to 20

In the first half of Module 5, students write numerals 0–5 and count to 20. They explore addition and subtraction stories with numbers 0–5, a natural way for them to understand *adding to* and *taking from*. Stories are acted out, modeled with objects, drawn, or solved using pictures. Children ask and answer questions about the story, such as "How many in all?" or "How many are left?" They learn to distinguish the question from the story.

Key Standards
- Know number names and the count sequence.
- Understand addition as *adding to*, and understand subtraction as *taking from*.
- Understand simple patterns.

Looking Back
In Module 4, students compared length, weight, capacity, and numbers to 5. Students also counted to 15.

Looking Ahead
In Topics D–F of Module 5, students will explore addition and subtraction stories using fingers, objects, and drawings. Students will also work with simple patterns.

(Above) Children learn to write numerals 0–5.

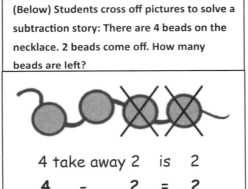

(Below) Students cross off pictures to solve a subtraction story: There are 4 beads on the necklace. 2 beads come off. How many beads are left?

How to Help at Home

- Make up addition or subtraction stories during everyday experiences. During bath time, say, "You have 3 toys in the tub. Here is 1 more toy. How many toys do you have now?"
- Work on a grocery list together. Have your child write the number of items needed, such as 5 apples, 2 boxes of cereal, or 1 carton of milk.
- Ask for help with counting during everyday experiences. While cooking, say, "I need 10 tomatoes. Can you count out 10 tomatoes for me?"
- When reading any book, have your child touch and count the number of objects in pictures: "1, 2, 3, 4, 5, 6, 7. There are 7 dogs!"
- Some lessons require additional materials. See your student's teacher for the additional materials required for each lesson.

Words and Key Terms
- Add
- Addition story
- Altogether
- Are left
- Equals
- In All
- Number sentence
- Plus
- Sixteen, seventeen, eighteen, nineteen, twenty
- Subtract
- Subtraction story
- Take away

A STORY OF UNITS

Family Math Newsletter PK•5

Spotlight on Writing Numerals

Throughout the Pre-K year, children have learned to identify numerals and match them to a number of objects. Now, their fine motor skills have developed to the point where most children are ready to write. The standard in Pre-K is writing 0–5, advancing to writing numerals 6–20 in Kindergarten.

Number Formation Chants

Simple rhymes help children remember how to write each numeral. Some children will say the rhyme each time they write until the strokes become automatic.

Curve from the top;
be a hero!
Close the loop, and
make a zero.

Top to bottom,
then I'm done.
I just wrote
the number 1.

Half a moon,
there's more to do;
slide to the right,
I wrote a 2.

Backward C,
backward C,
that is how
I write a 3.

Down the side,
to the right some more.
Top to bottom,
I've written 4.

Down the side,
around a hive.
Give it a hat.
I've written 5.

Focus on Tools: The Writing Rectangle

The writing rectangle is a tool to help children write numerals systematically, to make handwriting easier for them in the future. Writing rectangles use a dot to show where to start the numeral. If children start from the dot and keep the numeral inside the rectangle, they will not reverse their numbers as readily (i.e., write them backward).

Starting numbers and letters from the top is an important habit for your child to learn now. It will help her keep up when the writing demands increase in later grades. Numerals 1–5 all begin at the top left, with 0 starting in the center of the top side of the rectangle.

For some adults, it seems odd to start the 5 in the left corner and *add a hat* at the end. When in doubt, if children start at the left corner, they will be in the correct place for 7 out of 10 numerals (0, 8, and 9 are the exceptions). Starting the 5 at the left corner reinforces this idea.

At first, children will trace the numbers inside the writing rectangle with and without a writing instrument and then write them without tracing. Students do eventually write numerals 0–5 without the writing rectangle, but this tool provides them with a structure to form numerals correctly from the start.

Module 5: Addition and Subtraction Stories and Counting to 20

EUREKA MATH

Grade PK • Module 5 • Topics D–F
Family Math Newsletter

Addition and Subtraction Stories and Counting to 20

In the second half of Module 5, students continue to tell and solve addition and subtraction stories with numbers 0–5, now using fingers, cubes, math drawings, or numerals to represent the number of units (e.g., puppies) in the stories. For example, children solve, "Three puppies are playing. One puppy stops to rest. How many puppies are still playing?" using their fingers, cubes, or drawings of circles (see Spotlight on Math Strategies). In the final lessons, children replicate and extend patterns focusing on the repeating part of the pattern.

Key Standards
- Know number names and the count sequence.
- Understand addition as *adding to*, and understand subtraction as *taking from*.
- Understand simple patterns.

Looking Back
In the first half of Module 5, children learned to write numerals 0–5. They used actions, objects, and drawings to solve addition and subtraction stories.

Looking Ahead
In Kindergarten, children will begin the year building upon the sorting and classifying skills they learned in Pre-Kindergarten. They will count, write, and sequence numbers 0–10.

4 + 1 = 5

Join us for the Children's Math Theater to see our addition and subtraction stories come to life!

SAVE THE DATE

Words and Key Terms
- Add
- Addition story
- Altogether
- Are left
- Equals
- In all
- Math drawing
- Number sentence
- Pattern
- Repeating part
- Subtract
- Subtraction story
- Take away
- Total

How to Help at Home

- Make up addition or subtraction stories during everyday experiences. While grocery shopping, say, "There are 3 apples in the bag already. Let's put in 2 more. How many apples do we have now?" (This encourages your child to use his fingers to represent the story, since he can't see the apples in the bag.)
- Look for patterns as you move through your community. Children will see patterns in buildings, fences, clothing, and art.
- Build varied patterns with your child whenever possible. This encourages spatial reasoning. For example, when serving dinner, put the components in a certain pattern, and then replicate that pattern on the other plates.
- Some lessons require additional materials. See your student's teacher for the additional materials required for each lesson.

A STORY OF UNITS | Family Math Newsletter PK•5

Spotlight on Math Strategies

Drawings allow us to see mathematical situations and relationships in a way that helps make sense of the situation. The ability to represent a problem with a quick and abstract drawing will be key to children's math success throughout elementary school and beyond.

Sample Activity
(from Module 5, Lesson 25)

Dribble and Pass

Students count to 20 while practicing a fun pattern (dribble, pass, dribble, pass, …).

T: Let's use a dribble and pass pattern with imaginary basketballs.

T: First we'll dribble, and then we'll pass. Then we'll dribble, and then we'll…?

S: Pass!

T: You've got it! Now, let's count the Say Ten way as we dribble and pass.

T/S: 1 (dribble), 2 (pass), 3 (dribble), 4 (pass), 5 (dribble), …

In the lesson that follows, students learn to identify and duplicate patterns using objects.

Focus on Tools: Math Drawings

Math drawings are different from the drawings children create for artistic expression. In an artistic drawing, children may focus on details, color, or the type of media used. Math drawings focus only on representing the situation efficiently so that children can make sense of the situation and find an accurate solution promptly.

The drawings on the right represent the following addition story: Four lizards are running. Another lizard starts to run. Now, how many lizards are running? The image at the top shows a time-consuming, detailed drawing of the lizards. Below that is a math drawing using circles to represent the lizards (notice that the original lizards are shaded circles, and the new lizard is an empty circle). Another math drawing uses the letter *L* to represent lizards. Children can use any of these drawings to solve the problem, but the last two drawings took less time and effort to create, allowing the focus to stay on solving the problem.

By comparing their drawings with those of other students, children learn to think flexibly: "How are our drawings the same? How are they not exactly the same? Did we come to the same answer?" They see many different perspectives and make connections between them. As they learn more problem-solving strategies in later grades, this flexible thinking helps them continue to see and understand multiple ways to solve a problem.

152 | Module 5: | Addition and Subtraction Stories and Counting to 20

Copyright 2021 © Great Minds PBC

EUREKA MATH

Name _____ Date _____

How many in ⬛ ?

  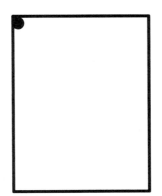 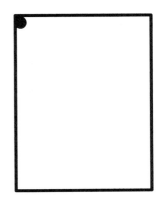

How many ?

  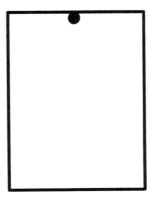 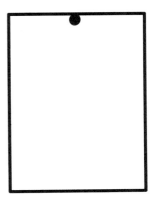

Lesson 1: Write numerals 0 and 1.

153

154

Name _____ Date _____

How many  ?

[2] [2] [ ] [ ]

How many  ?

[ ]

How many 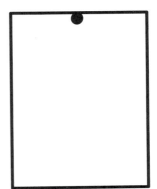 ?

[ ]

Lesson 2: Write numeral 2.

156

Name _____ Date _____

  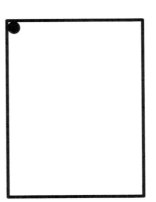 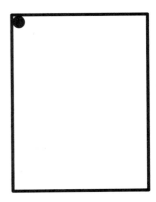

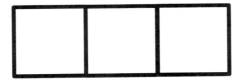

Lesson 3: Write numeral 3.

158

Name _____   Date _____

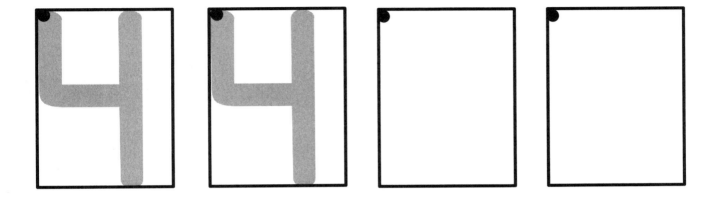

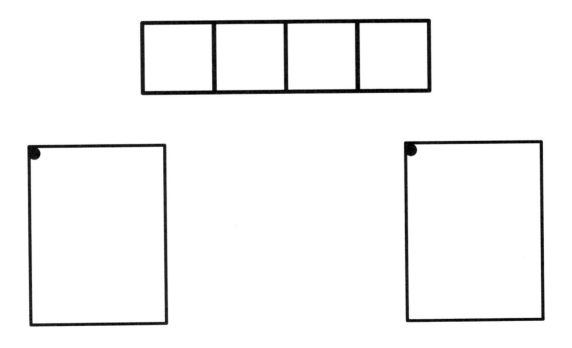

Lesson 4: Write numeral 4.

Menu

Pretzels

Strawberries

Juice

- -

Customer Order

Pretzels

Strawberries

Juice

menu items

Lesson 4: Write numeral 4.

Name _____   Date _____

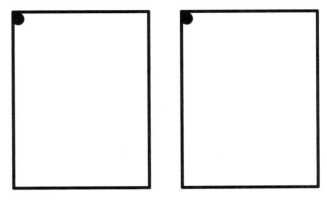

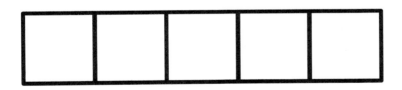

Lesson 5: Write numeral 5.

164

Menu

Pretzels

Strawberries

Juice

Customer Order

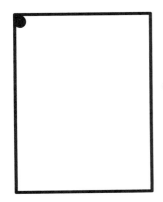
Pretzels

Strawberries

Juice

menu items

Lesson 5: Write numeral 5.

166

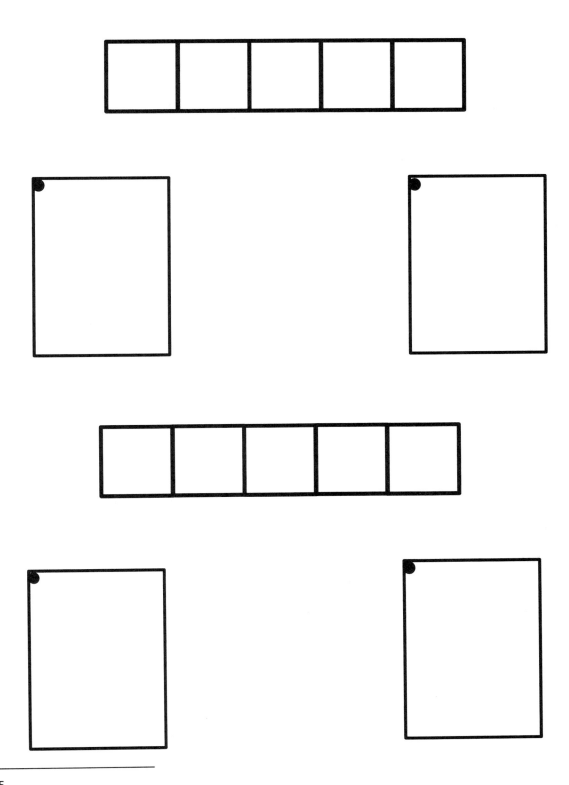

make 5

Lesson 5: Write numeral 5.

168

Cut along dashed lines to prepare the cards.

paper doll cards

170

Save this item for future use.

numeral writing rectangle

172

Cut along dashed lines to prepare the cards.

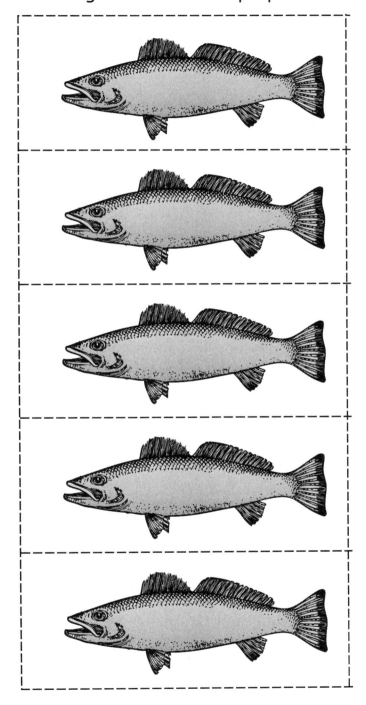

small fish cards

apple tree mat

Cut along dashed lines to prepare the cards.

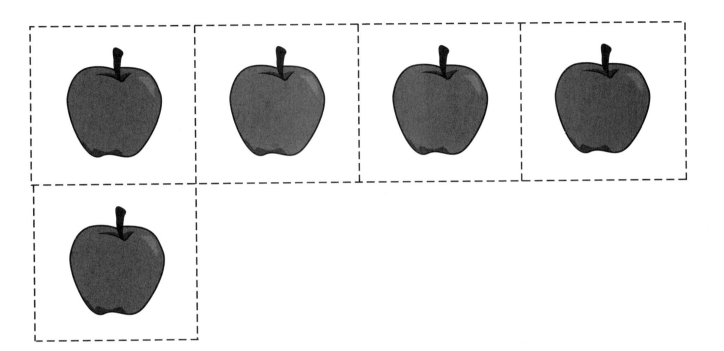

apple cards

178

Cut along dashed lines to prepare the cards.

paper doll cards

Name _____ Date _____

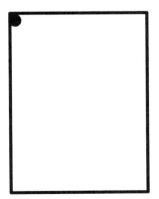

Lesson 13: Represent *take from with result unknown* story problems using number sentences.

182

A STORY OF UNITS

Lesson 14 Template PK•5

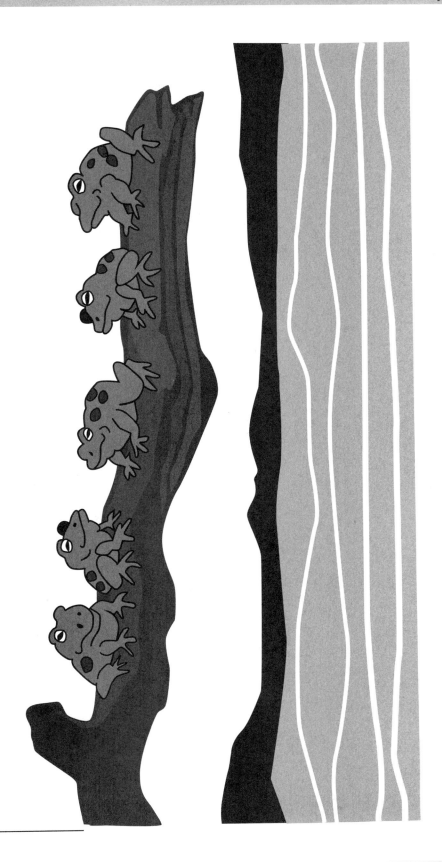

5 frogs picture

Lesson 14: Solve *take from with result unknown* story problems with objects from the story and drawings.

183

184

Assessment Template PK·5

Module 5: Addition and Subtraction Stories and Counting to 20

186

188

A STORY OF UNITS

Lesson 19 Template PK•5

_____

small writing rectangle

Lesson 19: Solve addition story problems with representative drawings.

A STORY OF UNITS Lesson 20 Fluency Template 1 PK•5

Cut along dashed lines to prepare the cards. Keep them for future use.

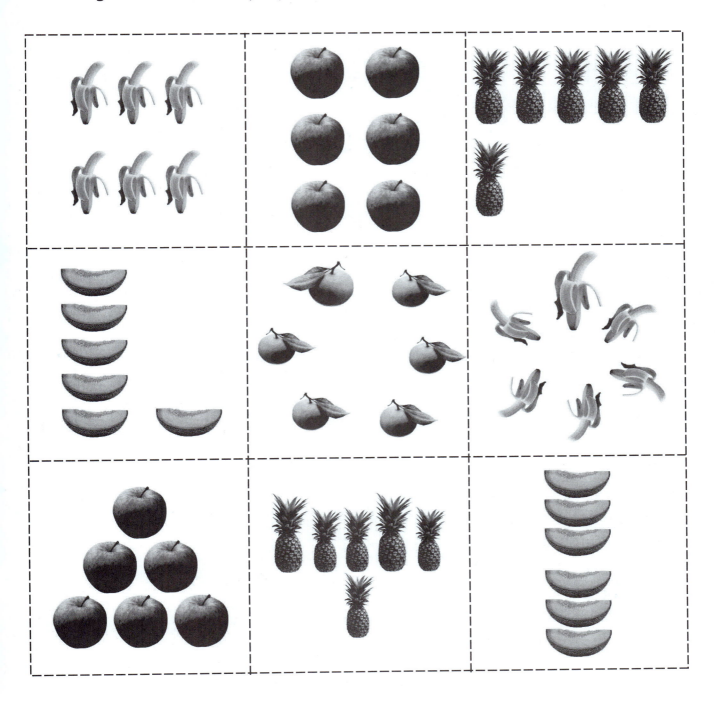

picture cards: 6

Lesson 20: Solve subtraction story problems using fingers.

191

192

A STORY OF UNITS   Lesson 20 Fluency Template 2   PK•5

Cut along dashed lines to prepare the cards. Keep them for future use.

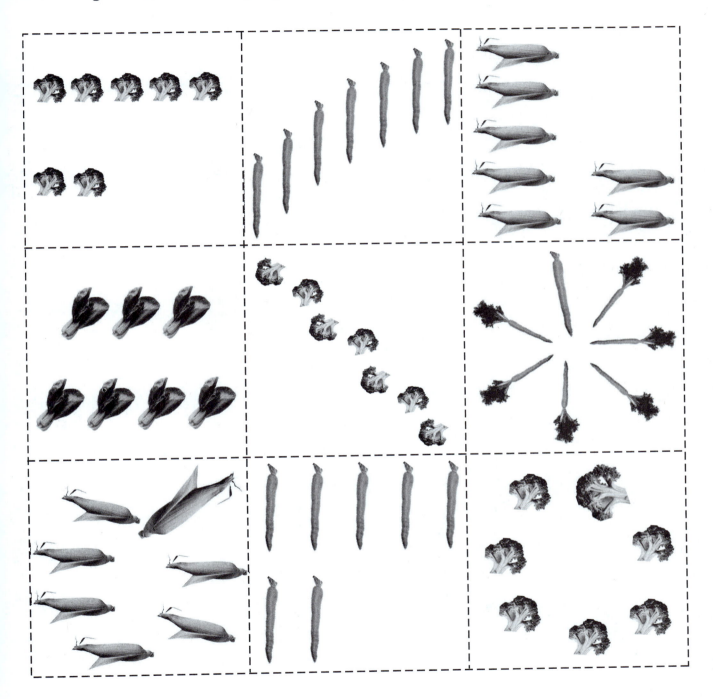

picture cards: 7

Lesson 20: Solve subtraction story problems using fingers.

194

Cut along dashed lines to prepare the cards. Keep them for future use.

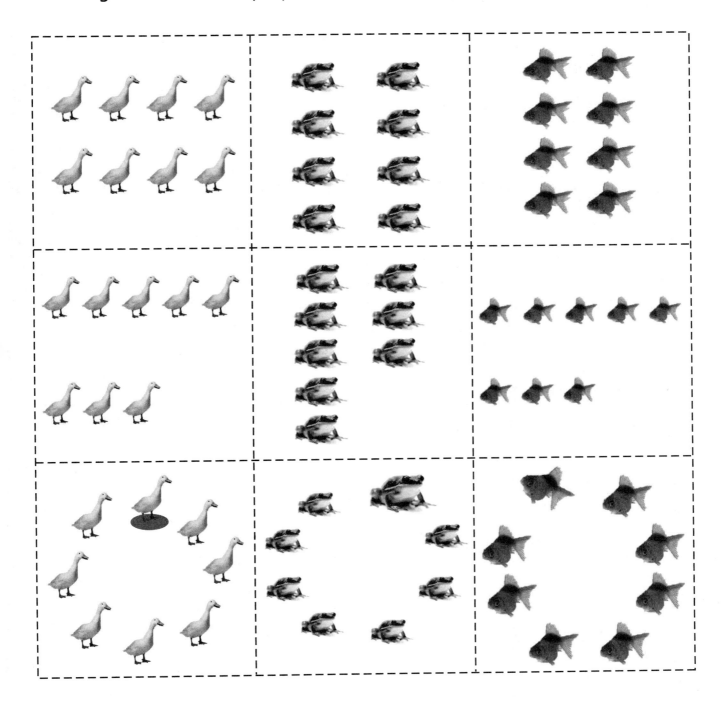

picture cards: 8

Lesson 21: Solve subtraction story problems using fingers.

196

A STORY OF UNITS
Lesson 22 Fluency Template PK•5

Cut along dashed lines to prepare the cards. Keep them for future use.

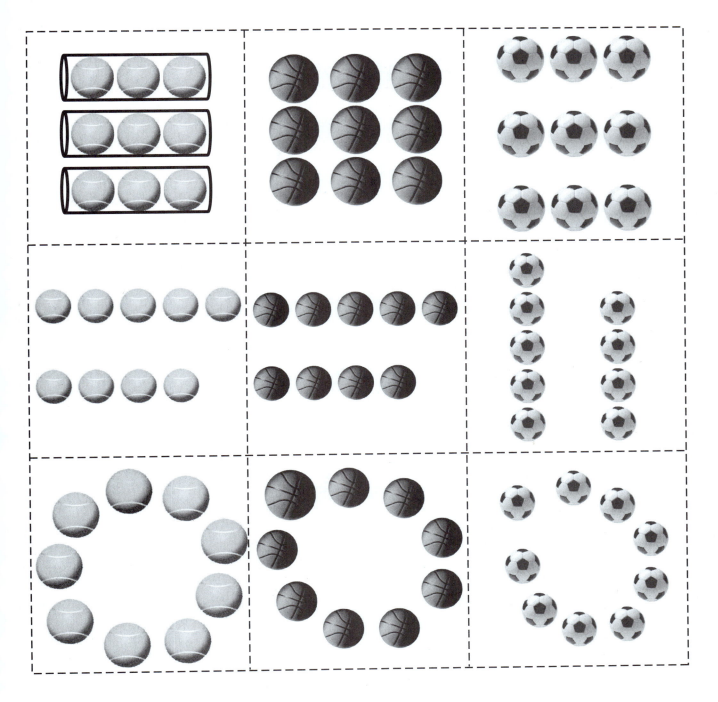

picture cards: 9

Lesson 22: Solve subtraction story problems with representative objects.

197

A STORY OF UNITS · Lesson 23 Fluency Template · PK•5

Cut along dashed lines to prepare the cards. Keep them for future use.

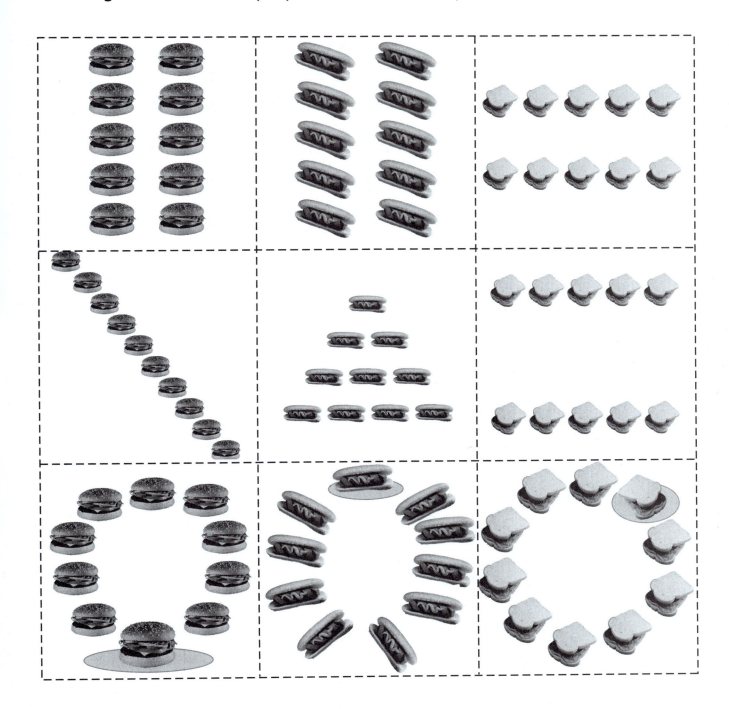

picture cards: 10

Lesson 23: Solve subtraction story problems with representative drawings.

199

Cut along dashed lines to prepare the cards.

sun and star cards

Lesson 24: Identify patterns using objects.

202

A STORY OF UNITS — Lesson 24 Template 2 PK•5

Cut along dashed lines to prepare the cards.

sun and star cards

Lesson 24: Identify patterns using objects.

204

Credits

Great Minds® has made every effort to obtain permission for the reprinting of all copyrighted material. If any owner of copyrighted material is not acknowledged herein, please contact Great Minds for proper acknowledgment in all future editions and reprints of this module.

p. 107 (from left to right) Photo by Andreas Weiland on Unsplash, Lightspring/Shutterstock.com, Becky Stares/Shutterstock.com, Fotosr52/Shutterstock.com, Svietlieisha Olena/Shutterstock.com, VectorShow/Shutterstock.com, Ylq/Shutterstock.com, Romolo Tavani/Shutterstock.com, Mega Pixel/Shutterstock.com, Metal Neck/Shutterstock.com, Photo by Oskars Sylwan on Unsplash p.113 Nerthuz/iStockphoto.com p. 139 Mike Pellinni/Shutterstock.com, dkingsleyfish/Shutterstock.com

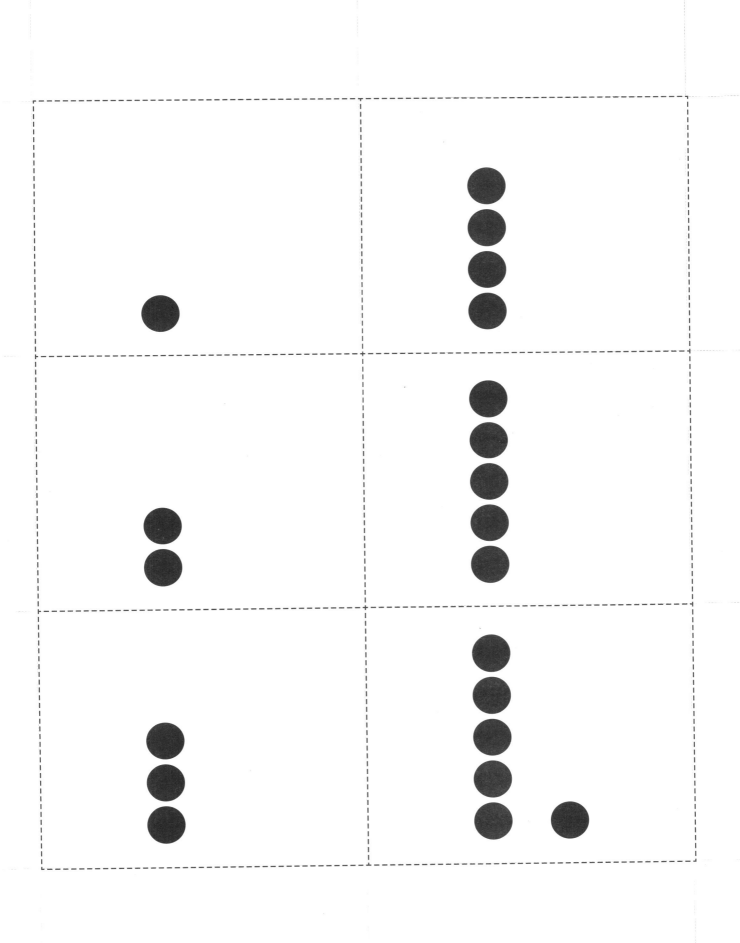

4   1

5   2

6   3

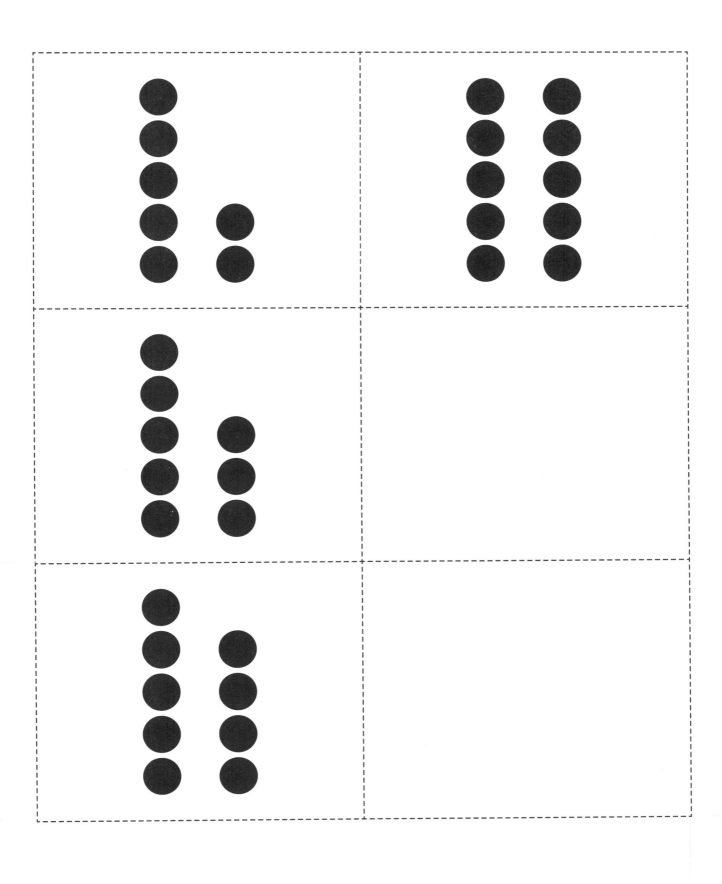

10

7

0

8

9